Effect of NaNO2 and

Panga Narasimha Reddy
Javed Ahmed Naqash
Avuthu Narender Reddy

Effect of NaNO2 and K2CO3 on Cold Weather Concrete

LAP LAMBERT Academic Publishing

Cover image: www.ingimage.com

Publisher:
LAP LAMBERT Academic Publishing
is a trademark of
International Book Market Service Ltd., member of OmniScriptum Publishing Group
17 Meldrum Street, Beau Bassin 71504, Mauritius

Printed at: see last page
ISBN: 978-613-9-45482-2

Effect of Sodium Nitrite and Potassium Carbonate on Cold Weather Concrete

Panga Narasimha Reddy
Javed Ahmed Naqash
Avuthu Narender Reddy
Bode Venkata Kavya Teja

Table of contents

List of Figures

List of Table

Abstract

Based on ACI 306R-10, the minimum temperature necessary for maintaining concrete hydration and strength gaining is 5 °C. If the temperature becomes lower than 5 °C, some special measures should be taken in order to prevent decrease in the rate of hydration and to prevent fresh concrete from freezing. Concreting in winter conditions is quite difficult. The reason is that concrete can harden and develop high strength in a relatively short time only within a certain range of above-zero temperatures. It is possible to create favorable conditions for concrete to harden when the ambient air temperature is below zero but that requires additional energy, materials and labor. Most of the cold weather countries spend annually plenty of money in order to facilitate concrete placing in the cold weather and to extend the construction season. During cold weather, preparations are made to protect the concrete; enclosures, windbreaks, portable heaters, insulated forms, and blankets are used to maintain the concrete temperature. As an alternative, antifreeze additives can be used in cold weather. These chemical additives suppress the freezing point of water under 0 °C degrees and accelerate the hydration of cement. The effects of antifreeze admixtures are often specific to certain cement brands and fine aggregates. It is important therefore to determine through laboratory mix trials both the operating range and the particular dosage of the admixture required for the intended application. The most common antifreeze admixtures are sodium nitrite ($NaNO_2$) and potassium carbonate (K_2CO_3). These antifreezes do not cause corrosion of steel and let concrete freeze at a temperature from −10°C ($NaNO_2$) to −15°C (K_2CO_3).

The objective of the study conducted was to optimize the proportions of two admixtures (Sodium nitrite, Potassium carbonate) to be used for cold weather concreting. The admixtures were tested for various properties, Compressive Strength, Tensile strength, Flexural Strength, Elastic modulus and Poisson's ratio. The samples were cast in two phases, in exterior winter conditions and under controlled conditions of -5°C with varying proportions of the admixtures. When compared to mixes without antifreeze admixtures the results showed a marked increase in the strength of the concrete samples.

CHAPTER 1

INTRODUCTION

1.1 GENARAL

Cold-weather conditions for construction is defined as a period when for more than three consecutive days, the following conditions exist: the average daily air temperature is less than 5°C (40°F) and the air temperature is not greater than 10°C (50°F) for more than one-half of any 24 h period. Under these conditions setting time and rate of strength gain of concrete is significantly delayed. Additionally, depending on the consistency of the mix, a reduced rate of hydration results in less water ingress into the cement particles, promoting bleeding and segregation. When the internal temperature of concrete decreased to -2°C (28°F), free water in the pores begins to crystallize as ice. Freezing increases the volume by 9% generating stresses that incorporate defects within the concrete When fresh concrete freezes, the strength of such concrete is lowered by 20–40%, its resistance to freeze-thaw cycling as given by the durability factor is lower by 40–60% and the bond between reinforcement and concrete is lowered by 70% compared with normally cured concrete. Thus, when concreting is done under cold weather conditions it is important to ensure that the concrete will not freeze while it is in the plastic state. Two options are available for cold weather concreting: (1) maintenance of near normal ambient and concrete temperatures through the heating of concrete ingredients and the provision of heated enclosures and (2) the use of chemical admixtures. Conventional non-chloride accelerating admixtures are used in cold weather concreting to offset the retarding effects of slow hydration on the rate of strength development. Such admixtures however do not permit concreting at or below freezing temperatures. When concreting is carried out under more drastic weather conditions, special admixtures, called antifreeze admixtures, which affect the physical condition of the mix water, are used. Antifreeze admixtures are capable of depressing the freezing point of water in concrete considerably and their use at temperatures as low as -30°C enables an extension of the period of construction activity. Although the use of antifreeze admixtures has been an acceptable practice in Russia for nearly three decades, their use in other countries has been more recent. The former contain salts of formates, nitrates and nitrites and are effective for set acceleration and strength development. However, their effectiveness is dependent on the ambient temperature at the time of placement. Chemicals used as antifreeze admixtures include sodium and calcium chloride, potash, sodium nitrite, calcium nitrate, urea, and binary systems such as calcium nitrite-nitrate and calcium chloride-nitrite-nitrate.

1.2 BACKGROUND

Concreting during cold weather poses various challenges in terms of placement of concrete, strength gain, curing and so on. Behavior of concrete hydration changes with regards to temperature, requiring different heat inducing techniques for satisfactory placement at low temperatures. Employment of such techniques proves to be uneconomical. The cold weather experienced in certain areas for a considerable amount of time in a year reduces the construction season drastically. The need to tackle the problem of reduced construction season has led to efforts of developing a concrete that is efficient at low temperatures.

Solutions such as heating enclosures and insulation have been applied to the problem of cold weather concreting since the 1930s. However these methods have no undergone any considerable amount of change until recently. Heating enclosures and insulation techniques are extremely costly, consume a huge amount of energy and also require skilled labor. The recent work focuses on the development of antifreeze admixtures to depress the freezing point of water thereby making it possible for the concrete to achieve maximum possible strength at low ambient temperatures. This reduces the cost of construction and energy required considerably.

1.3 PROBLEM STATEMENT

As mentioned in Section 1.1 the recent work in the area of cold weather concreting is focused on the use of antifreeze admixtures for the placement of concrete at low temperatures in accordance with the specifications of ACI 306-02. Admixtures such as $NaNO_2$, K_2CO_3 etc. have been tested individually as possible solutions to the problem of cold weather concreting. Admixtures have also been tested in combination with each other. However the proportions of these admixtures were either as prescribed by the manufacturer or as mentioned in the ASTM 494 C. The current cold weather concreting code ACI 306-02 does not provide any specific details about the methods by which the standard atmospheric conditions for concreting must be maintained at low ambient temperatures. Similarly ASTM 494 C, the standard for specifications of chemical admixtures for concrete, provides limiting values for the proportions of individual chemical admixtures to be used in concrete. However, it does not provide any clear idea about the recommended proportions for the admixtures.

Also the effect of various antifreeze admixtures on the compressive strength has been previously investigated by researchers. However little has been done to study the effect of antifreeze admixtures on the elastic properties of concrete. The modulus of elasticity is essentially a measurement of the stiffness of a material. Modulus of elasticity of concrete is a key factor for estimating the deformation of buildings and members, as well as a fundamental factor for determining modular ratio, m, which is used for the design of section of members subjected to flexure. The precise determination of the modulus of elasticity of concrete is very important for structures that require strict control of the deformability.

1.4 OBJECTIVES

The Objectives of the present study are:

- To study the effect of antifreeze admixtures (sodium nitrite and potassium carbonate) on the strength and elastic properties of concrete produced in exterior winter conditions in Jammu & Kashmir.
- To study the effect of antifreeze admixtures (sodium nitrite and potassium carbonate) on the properties of concrete under controlled conditions.
- To suggest an antifreeze admixture and optimize its dosage for use during winter.

The strength properties include the study of compressive, tensile and flexural strength according to IS 516-1959. The modulus of elasticity and Poisson's ratio are to be determined by an arrangement of a compressometer and extensometer according to ASTM C 469-02.

CHAPTER 2

REVIEW OF LITERATURE

American Concrete Institute ACI-212 (ACI 1985) specifically prohibits antifreeze admixtures. Antifreeze admixtures are not currently included as a cold weather concreting option in that report, although consideration is being given to including their mention in a future update. Section 3.2.2 of that report states, "No materials are known which will substantially lower the freezing point of the water in concrete without being harmful to the concrete in other respects (p. 212.1 R-7)" however a new section, ASTM C09.23.5, has recently been created to oversee the development of an appropriate test method and specification for this class of chemical admixtures.

2.1 MODERN VIEWS ON HARDENING OF CONCRETE

The hardening of concrete, whereby a plastic mass turns into an artificial stone that has good physio-mechanical characteristics, is accompanied by a set of very complex phenomena that have not been studied well enough yet and are not fully controllable. The reasons are the versatility of active ingredients of the original binder (cement), the complexity of the system of new formations, and the multiplicity of components in the material as a whole, which include substances in different states of aggregation. Due to relaxation with time, the microstructure and macrostructure of concrete varies and it becomes a new system with different properties at each hardening step.

The processes of chemical and physical transformations slow down considerably as temperature drops and go faster as it rises. So temperature factor is regarded as a most effective influence on the hardening of concrete.

The concrete hardening processes were studied by many researchers, including A.A.Baikov, P.P.Budnikov, A.E. Sheikin, G.Green Kenneth, T. Powers and others. Their points of view on the hardening differ greatly and combined into one theory. However the crystallization theory of hardening has an overwhelming number of supporters. According to the crystallization theory of hardening, minerals of the cement clinker are dissolved and interact with water as cement is mixed with it. The liquid phase is rapidly saturated with calcium hydroxide formed by hydration of C3S. Gypsum and alkalis contained in the clinker are quickly dissolved.

The saturation of liquid phase with calcium hydroxide is very fast because it requires only dissolving CaO that is formed when 0.3 to 1% of C_3S contained in the cement reacted with water. In fact, more CaO goes into the liquid phase and it forms a supersaturated solution since 1 to 4% of C3S hydrates within first three minutes after cement is mixed with water even at normal temperature. Thus, the hydration of minerals of the Portland cement clinker takes place in saturated solutions of $Ca(OH)_2$ rich in caustic alkalis and gypsum. As they hydrate, calcium silicates, C3S and C2S, form calcium hydrosilicates and calcium hydroxide. The latter precipitates in the form of prisms that are gradually transformed into lamellar crystals. The calcium hydrosilicates undergo changes that involve transformation into a high or low basic modification depending on concentration of CaO in the solution. Tricalcium aluminate (C3A) reacts with gypsum to form a high sulphate modification of calcium hydrosulfoaluminate. Tetracalcium aluminate ferrite (C4AF) hydrates and some part of it also react with gypsum producing the high sulfate form of calcium hydrosulfoaluminate and iron-containing phase. In addition to the minerals described above. Portland cement clinker contains the glass-forming phase that consists of uncrystallized C3S and C4AF with inclusions of CaO and SiO2. When Portland cement is mixed with water, cement phase hydrates and forms hydroaluminates, hydroaluminate ferrites, and hydrogranates. The latter crystallize in the form of regular isometric crystals. New formations produced by the hydration of cement make up the structure of a new material, i.e., hardened cement paste. The formation of the structure is a complex process that has not been studied completely. The mechanism of cement hydration is the central question there and it can be represented on the basis of modern views as follows.

Mixing cement with water triggers dissolution and hydration of clinker minerals. The liquid phase is saturated and, when oversaturation with hydration products is reached, newly formed hydrates begin to crystallize in the form of submicrocrystals. The higher the oversaturation of a solution the faster the crystallization process with the appearance of more disperses particles. Spreading over the whole system, the crystallohydrates form a coagulation structure. As the number of crystals grows, they start to move closer together. Mechanical admixtures, quartz sand in particular, have a favorable impact at this stage of hardening. Acting as an active backing, they help to accelerate the crystallization, to dissolve the initial binder, and to reduce the probability that the nuclei of a new phase will be formed on the surface of the latter's grains. Thus, the presence of the quartz aggregate should weaken the isolation of cement grains and increase the degree of hydration postponing the retardation or discontinuance of the hydration until later. The acceleration of crystallization and hardening at the boundary with the aggregate is accompanied by an increase in the strength of interface layers of the hardened paste. Strong crystallizing contacts of intergrowth are formed between individual small crystals of hydrates in the process of crystallization. Not all crystals hydrates in the hydrated cement paste, however, can form regular intergrowths. The priority here is given to crystal hydrates of the same structural type. It appears that intergrowths of hydrosilicates and hydroaluminates lead an independent existence in the mass of the hardened cement paste. Besides, the structure carries a great number of individual hydrate crystals and their aggregations whose individual particles intertwine and are held together by mechanical forces of adhesion. Intergrowth contacts may arise only after many particles of new formations appear in the system and their spacing does not exceed the double thickness of the adsorption layer of the hydrate molecules. The crystals join into individual aggregations that intergrow to form the crystal skeleton of the hardened cement paste, a solid artificial polymineral. Then new crystals grow around the skeleton filling its voids and reinforcing it. Gradually a system with a weak coagulation structure turns into a crystalline structure. The amount of gel decreases because the particles grow and join each other and also because water is sucked to the nuclei of the cement particles which have not hydrated yet. The system becomes stronger and more rigid. The spontaneous crystallization of hydrate forms gives rise to thermodynamically unbalanced contacts that distort the crystal lattice of the material. So later, when concrete is cured in wet conditions, there is recrystallization that starts by itself and results in dissolution of thermodynamically unstable compounds and in growth of regularly formed crystals. The recrystallization weakens the structures at certain stages and reduces the strength of the hardened cement paste. So the strength of the hardened paste depends on the growth of small crystals, their joining into a rigid skeleton, and the reinforcement of the skeleton by new small crystals that grow on it. Any recrystallization in an intergrowth already formed can only weaken its structure. The growth of the crystal skeleton not only increases strength but also causes internal stresses due to directed growth of crystals that are already bound by intergrowth contacts. The internal stresses (similar to the crystallization pressure) weaken the system because they partially destroy it in the weakest sections. The strength of Portland cement usually increases in hardening because the destructive process is more than compensated by constructive process of formation of new intergrowth contacts and an increase in crystal sizes.

An increase in temperature does not affect the quality of the hardening mechanism of the Portland cement binder but increases the rate of the structuring processes (accelerated dissolution of the initial material, fast growth of smaller crystals in new formations, etc.). Imposition of vibration effects at the stage of the coagulation structure helps to bring together and to pack more closely the particles and to form close coagulation bonds necessary to form the crystalline structure.

According to A.F.Polak, a close coagulation structure is 10 to 100 times stronger than that with a widely spaced coagulation, i.e., when the distance between particles exceeds the double thickness of the adsorption layer that consists of hydrate molecules. Thin films of hydrosulfoaluminates and calcium hydrosilicates formed on the surface of hydrating cement grains at the start of the process can be easily penetrated by water that diffuses from the intergrain space and from the saturated liquid phase which enters the intergrain space. As the films thicken and condense, the diffusion of ions through it becomes more difficult with time. The liquid phase in the space between the colloidal film and the cement grain surface (transition zone) becomes more difficult to penetrate and more saturated as compared with the liquid phase in the capillary space. The difference in concentrations causes osmotic pressure. The gel-like films dissolve by themselves after a time and are destroyed by osmotic forces and by the increase in volume of new formations that crystallize on the internal surfaces of the films. So the reaction of the cement with water is accelerated again. When the temperature is high, films formed around cement grains at the early stage of hardening deteriorate very quickly and the unreacted parts of the grains become exposed again to moisture. This phenomenon agrees well with the nature of heat evolution that has extremums at early stages, which correspond to the initial period of hydration and to its new increase after the films are destroyed. As the hardening progresses, the hydration processes slow down considerably and become very weak after 28 days of curing in favorable heat and moisture conditions. Strength properties of concrete may decrease and increase again from time to time as a result of the recrystallization processes and accumulated internal stresses. As mentioned above, an increase in temperature affects greatly hardening process and formation of the microstructure and macrostructure of concrete. Lower viscosity and higher reactivity of water due to heating help to accelerate the dissolution of the binder, to oversaturate the liquid phase, and to produce a great number of crystallization nuclei and a finer dispersed structure. This in turn reduces internal stresses under optimum conditions of thermal treatment. An accelerated temperature rise that may be produced by electric contact heating allows to destroy the unstable films around hydrating cement grains and to achieve practically the same degree of hydration as in mild conditions of thermal treatment. A higher hardening temperature helps to turn unstable compounds into more stable ones (e.g., hexagonal crystals of hydroaluminates into crystals of hexahydrate tricalcium aluminate and high basic calcium hydrosilicates into basic ones). So the thermal treatment can shorten the time of physio-chemical processes in concrete as compared with hardening under normal conditions.

The strength characteristics of concrete are known to depend not only on the strength of the hardened cement paste structure but also on the structure of the system as a whole and on reliability of bonds between components under external loading. From this point of view, the structural and physical phenomena observed in concrete under thermal treatment are quite important. Heating of concrete as of any physical body results in expansion of its constituents that have different thermal properties. The expansion coefficients of the concretes solid components do not differ much but those of water and entrapped air differ from the expansion coefficients of the solid components by one and two orders of magnitude, respectively. A temperature of 60°C or higher accelerates internal evaporation of moisture in the capillary-pore system (i.e., concrete), which increase the content of the vapor-air mixture in the material and thus its internal pressure, loosens the concrete, and decreases its density and strength.

This very short analysis of the concrete structuring process in hardening shows that heating of fresh concrete creates on the whole favorable conditions for accelerated transformation of the material into an artificial conglomerate stone. The chemistry of hardening remains practically the same under thermal treatment but the reactions go much faster. The temperature factor affects the microstructure of concrete and causes certain defects which may adversely affect the properties of concrete and the durability of structural members made of it. Certain techniques make it possible either to avoid the structural defects or to reduce them substantially.

So, thermal treatment is an effective method for accelerating the hardening of concrete under various temperature conditions of the environment.

2.2 THE MECHANISM OF DETERIORATION OF CONCRETE DUE TO EARLY FREEZING

The effects of frost on concrete have been studied by many researchers prominent among them are V.M.Moskvin, S.A.Mirionov, and I.A.Kireenko. Studies that where conducted at NIIZhB (Research Institute of Concrete And Reinforced Concrete) in Russia found a number of irregularities in the destructive phenomenon that arises in concrete when it freezes, thaws and then hardens at and above zero temperature. Causes of deterioration of concrete can be divided into following three groups:

- Conversion of water into ice, that causes an increase in volume and great internal pressure.
- Mass transfer as the concrete freezes, which leads to the distribution of moisture and the formation of large ice segregations.
- Disintegration of the whole multi component system due to mass transfer and sedimentation phenomena, which weakens the bond between the coarse aggregate particles and the paste constituents of the concrete.

2.2.1 DETERIORATION OF CONCRETE IN FREEZING

The conversion of water into ice increases the volume by 9% that causes high internal pressures. When frozen immediately after mixing, concrete has no elasticity and cannot resist the frozen moisture that fills its intergrain space. This pressure forces the solid particles apart. A specimen contracts as it is cooled down to 0°C. There is an abrupt change in strains within the temperature range between 0 and -1°C, which characterizes expansion of the concrete due to conversion of mechanically bound moisture into solid phase. The strains gradually decrease as the concrete is cooled further. So in fresh concrete, at temperatures from 0 to -2°C, most of the moisture (upto 90%) turns into ice and causes a high internal pressure in the material, which cannot be resisted by the plastic system without deformation and an increase in volume.

When the temperature of frozen concrete rises, its volume increases again due to expansion of all its constituents. As the temperature crosses the zero point and ice starts to melt, the internal pressure drops as a result of the contraction of the liquid volume and of the opening of channels for migration of the gaseous fraction that was blocked in the voids by ice. The loose system starts to harden quickly when the temperature is above 0°C. The strength of such concrete remains lower than that of unfrozen concrete that hardened at normal temperature and humidity. The expansion of water in freezing also damages concretes that had attained some strength by the time it freezes. However, as concrete hardens its elastic characteristics improve, the damage done to its structure by freezing diminish. The reason in the reduction is the amount of mechanically bound water which noticeably affects strains. Thus concrete frozen 14 to 17 hours after hardening under normal conditions has no strains when the temperature goes above 0°C. Changing porosity of the material also helps. Macropores that get rid of mechanically bound water as it is bound chemically and physically or evaporates are filled with air and become reservoirs where the air compresses when the liquid phase freezes and so the capillary walls are not ruptured to any great extent.

2.2.2 STRUCTURAL DETERIORATION DUE TO MASS TRANSFER

Mass transfer causes great damage to the concrete structure primarily of physical nature .When concrete is mixed in a mixer; water spreads evenly over the whole mass. Moisture may redistribute during transportation, handling, and storage of mixed concrete. So some parts of the mixture may become excessively saturated with water and large structural faults can be expected there in freezing. But even if the moisture was distributed uniformly over the concrete mass after placing and compaction, it starts to move after freezing and the appearance of a temperature gradient in the structural member.

The member starts to cool from the surface of concrete, with the cold front gradually spreading into the inner core. Since partial pressure is lower in the area of lower temperatures, the liquid phase starts to migrate to the colder area. Having reached the zero isotherm, it freezes and expands forcing the grains of cement and aggregate apart in fresh concrete or rupturing the crystal skeleton in the concrete that began to harden. Ice inclusions are formed in the pores and cavities. Water diffuses to the ice inclusions from thinner capillaries (including those inside gel) because these inclusions are cold sinks inside the material. Another reason is that the reaction of cement with water is exothermal and involves heat evolution. Although frozen concrete contains little water and the latter has low reactivity, still the reaction does take place and some exothermal heat evolves raising the temperature on the surface of cement grains. This causes the moisture to migrate to the colder area that is to the ice inclusions, where it changes not mechanically bound water from adsorptionally bound one and freezes. The ice inclusions increase in volume and gradually from ice segregations which are often visible, particularly in the layers close to the concrete surface. The migration of water in hardening and hardened cement mortars or concretes may cause failure of a structure when the temperature gradient is high.

An example is splitting of glass blocks in a window opening during construction of a market place at city in Siberia. Ice growths appeared from the outside of the building in mortar joints between glass blocks. The growths built up rapidly in the frost that reached -45°C. Inspection of the facility and an analysis of the splitting have shown that the reason was migration of moisture. Humidity in the room was high when it was being finished. The temperature drop between the air inside and outside was 60 to 65°C.The mortar in the joints 10 to 15 mm wide and

150 mm deep (the thickness of a glass block) was placed in summer and must have been highly porous. In winter, the temperature of the part of each joint adjacent to the interior surface of the window was above zero and of that adjacent to the exterior surface, below zero. The zero isotherm passed at about 1/3 of the distance from the exterior surface and the mortar in the joints acts as a kind of a pump that delivered moisture from the warm area to the cold one. There the moisture froze and expanded causing high internal stresses in the mortar, which cut off the exterior halves of the glass blocks. Internal glazing is recommended to prevent excess of the moisture in the room to the mortar joints and to stop this destructive process.

2.2.3 DISRUPTION OF THE SOLIDITY OF CONCRETE

When the fresh concrete freezes, the interaction between the coarse aggregate particles and paste is impaired by the mass transfer. The particles of coarse aggregate in normal weight concrete are usually denser and conduct heat better than the cement paste. So their temperature drops down faster when the concrete cools and they attract moisture from the core. As it freezes with the interface with the aggregates, the water forms an ice film that continues to grow due to diffusion of the moisture from capillaries. The film may be 1 mm thick or more.

The ice film at the interface of the paste and of the coarse aggregate particles benefits sedimentation, too. The ice film originally formed by sedimentation may also grow under favorable conditions due to the influx of moisture from capillaries. As it grows in volume, the ice filling widens the gap between the aggregates and the paste not only breaking their contact but also losing the structure of the concrete in general. The film disappears when it melts and the resultant air gaps disrupt the adhesion of the components.

The situation is different when frozen concrete contains porous aggregates. The aggregate particles absorb moisture and no water film is formed at their interface with the cement paste as a result. So the concrete deteriorates less. Porous aggregates, such as expanded clay, pumice, or volcanic slag, can reduce considerably the frost damage to concrete at an early age. The absorption of a part of moisture from cement paste by particles with a multitude of microcapillaries and pores before freezing reduces the water-cement ratio. The amount of mechanically bound water in the paste falls shortly. When the concrete freezes, the water remaining in the intergrain space does not disrupt the structure too much because the expansion of the moisture is compensated by a great number of air-filled pores.

It should be emphasized, however, that in concretes with a high W/C ratio the porous aggregate increases frost damage rather than prevents it since the aggregates is highly saturated with moisture, its particles break when they freeze and the quality of the concrete deteriorates dramatically.

2.3 EFFECTS OF ANTIFREEZE ADMIXTURES ON CONCRETE

As with any additive, there is concern over antifreeze admixtures' effects on concrete properties. As can be seen, the rate of strength development for low-temperature concrete made with antifreeze admixtures lags that of similar room-temperature concrete made without admixtures. This strength lag varies with admixture and with temperature. For example, strength gains range from a high of 93% of the 28-day strength of control samples for ammonia at -20°C to a low of 29% for calcium nitrate plus urea at -10°C.

2.3.1 COMPRESSIVE STRENGTH

A common indicator of concrete quality is compressive strength. Safe and economical scheduling of crucial construction operations makes it important that concrete attains certain strength before work progresses. Compressive strengths are shown in Tables 1 and 2. In tests conducted for as long as 1-1/2 years, the Low

Temperature Building Sciences Institute of China (1979) showed that concrete made with sodium nitrite plus sodium sulphate gained 114% of 28- day design strength. In practice, when air temperatures rise, concrete strength gain should accelerate. Kivekas et a1. (1985) provide an example of this temperature effect. They showed the typical early age strength lag for a variety of admixtures cured at low temperature; however, when the admixtures are cured at room temperature for an additional period, strengths show a marked improvement. Further, when the water-cement ratio (w/c) of the sodium nitrite plus calcium chloride concrete mix was reduced to 0.59 from 0.66 by including a water reducer, compressive strength improved. This has practical implications as concrete placed in the winter cannot only be expected to slowly gain strength but can be expected to gain appreciable strength the ensuing spring and summer, provided, of course, that proper moisture conditions are maintained. Fatma Karagöl, Ramazan Demirboğa, Mehmet Akif Kaygusuz, Mehrzad Mohabbi Yadollahi, Rıza Polat in "The influence of calcium nitrate as antifreeze admixture on the compressive strength of concrete exposed to low temperatures (2012)" showed that Compressive strength of both control and calcium nitrate concrete was decreased with decreasing deepfreeze temperatures. However, reductions in strength of concrete with calcium nitrate were lower than that of control sample. Calcium nitrate concrete's compressive strength was high enough up to 14 days of exposure of −10 °C deepfreeze temperature without any additional precautions. However, after 28 days of deepfreeze curing it can conserve concrete that exposed to −5 °C only without any protection or additional water curing. Fatma Karagol , Ramazan Demirboga , Waleed H. Khushefati in "Behavior of fresh and hardened concretes with antifreeze admixtures in deep-freeze low temperatures and exterior winter conditions" showed an increment of 108% in compressive strength of sample with 4.5% calcium nitrate and 4.5% urea when compared to the compressive strength of the control sample exposed to the same conditions.

2.3.2 TENSILE STRENGTH

Another important design parameter is the tensile strength of concrete. It is particularly important for pavements. Tensile strength is also important in determining how well concrete withstands frost. Those concretes with high tensile strengths are better able to resist the expansive stresses that arise in the process of ice formation. Unfortunately, very little is published on this concrete property. In the little that is published, Goncharova and Ivanov (1975) state that tensile, compressive and cohesive strengths are "on average the same as those of additive free concrete hardened under normal humidity conditions," while Grapp et al. (1975) show that the engineering properties of concrete were affected by potash, the one admixture they tested. According to Grapp et al., there is a measured decrease in the dynamic modulus of elasticity, an increase in the coefficient of expansion and a decrease in the split-tensile strength of concrete containing potash.

2.3.3 COST EFFECTIVENESS OF ANTIFREEZE ADMIXTURES

Winter increases construction costs. Much of this increase is caused by a drop in the efficiency of construction machinery and manual labour. Jokela et al. (1982) note that machinery costs increase 1.6 to 2 times and manual labour takes about twice as long when work is done in cold weather. Some estimates have placed winter construction costs even higher. According to Mironov and Demidov (1978), winter conditions can increase concreting costs by a factor of 1.5-2 High heating requirements and the need for protection account for much of this increase. When using concretes containing antifreeze admixtures, however, the need for special protection diminishes and any increased costs associated with the antifreeze admixtures are offset by the reduced winter protection requirements. Kuzmin (1976) says that up to 94% of the added cost of using antifreeze admixtures is attributable to the cost of the admixture itself. The remaining 6% is associated with processing and handling the admixture. Therefore, the cost effectiveness of antifreeze admixtures can be reasonably estimated on the basis of admixture costs.

Antifreeze admixtures are mentioned in International Union of Testing and Research Laboratories for Materials and Structures (RILEM) publication on cold weather concreting (Kukko and Koskinen). A section of that guide, devoted to antifreeze admixtures, provided a table of seven antifreeze admixtures with their low-temperature strengths. That table, shown as Table 1 in this report, lists the strengths of concrete maintained at the temperatures shown. This table, however, is only intended as a guide. RILEM recommends that concrete containing any antifreeze admixture be tested in the laboratory before being used in the field. Probably because of that caution, RILEM does not give mix proportions that would allow one to apply Table 1 directly in the field without first conducting laboratory experiments.

To get an idea of what antifreeze concentrations might be necessary to achieve the results shown in Table I, Table 2 was developed from the literature. Data on strength, temperature and admixture concentration were selected from publications that most closely agreed with the strength and temperature data in Table 1. For those situations where strengths were not available, admixture concentrations and service temperatures were still provided to indicate potential ranges of use. In addition, Table 2 lists helpful references on each admixture.

Admixture composition	Mean Temp of concrete hardening (°C)	Strength % (of control sample cured at 28°C)			
		7days	14days	28days	90days
100% $NaNO_2$	-5	30	50	70	90
75% $Ca(N0_3)_2$ + 25% $CO(NH_2)_2$	-10	20	35	55	70
	-15	10	25	35	50
100% NaCl	-5	35	65	80	100
70% NaCl + 30% $CaCl_2$	-10	25	35	45	70
40% NaCl + 60% $CaCl_2$	-15	15	25	35	50
50% $NaN0_2$ + 50% $CaCl_2$	-5	40	60	80	100
	-10	25	40	50	80
	-15	20	35	45	70
	-20	15	30	40	60
100% K_2CO_3	-5	50	65	75	100
	-10	30	50	70	90
	-15	25	40	65	80
	-20	25	40	55	70
	-25	20	30	50	60

Table 2.1 RILEM recommendations for cold weather concreting with antifreeze admixtures (after Kukko and Koskinen 1988)

Chemical	% By Weight Of Cement	Strength* (%)	Temperature (°C)	References
$CaCl_2$	7	50	-15	a,g,j,m,n,o,p,q,r
NaCl	5.7	80	-5	a,b,c,j,m,n,p,q
$NaN0_2$	6	70	-5	a,d,e,s
	8	57	-10	a,d,e,s
	10	36	-20	a,d,e,s
NaCl + $CaCl_2$	7.7	58	-20	g,h,n,t,u,v
$CaCl_2$ + $NaNO_2$	5		-5	b,c,e,h
	6.5	-10	14	b,c,e,h
	8.5	-15	5	b,c,e,h
	9	42	-20	b,c,e,h
$Ca(N0_2)_2$ + $CO(NH_2)_2$	5.5	-5	23	b,c,f,k,l,w
	9.5	55	-10	b,c,f,k,l,w
	11	35	-15	b,c,f,k,l,w
	13	-20	-4	b,c,f,k,l,w
$Ca(N0_3)_2$ + $CO(NH_2)_2$	8.8	29	-10	e,f,h,w
	9	34	-20	e,f,h,w
$Ca(N0_3)_2$ + $NaSO_4$	6.6	56	-10	e,f,h,w
$Ca(N0_2)/(N0_3)_2$ + $CaCl_2$ + $CO(NH_2)$	9	61	-20	f,h,w
	11.5	40	-20	f,h,w
	13	35	-20	f,h,w
	14	20	-25	b,c
K_2CO_3	6	75	-5	b,c,d
	8	70	-10	b,c,f,i
	10	65	-15	b,c,f,i
	10	47	-20	b,c,f,i
	12	55	-20	b,c,f,i

Table 2.2 Common antifreeze admixtures for cold weather concreting

* Percent strength of control sample cured at room temperature for 28 days . a-Mironov et al. (1976); b- Kpkko and Koskinen (1988); c-Jokela et al. (1982); d-Low Temperature Building Sciences Institute (1979); e-Kivekas et al. (1985); f-Krylov et al. (1979); g-Mironov (1977); h-Goncharova and Ivanov (1975); i-Grapp et al. (1975); j-Kuzmin (1976); k-Virmani et al. (1983); 1-Virmani (1988); m-Derrington (1967); n-Stormer (1970); o-Kostyayev et al. (1971); p--Yang (1982); q-Cottringer and Kendall (1923); r-Yates (1941); s-Mironov et al. (1979); t-Miettinen et al. (1981); u-Mironov and Krylov (1956); vizov (1956); w-Golobov et al. (1974); x-Kuz'min et al. (1976); y-Bazhenov et al. (1974).

CHAPTER 3

EXPERIMENTAL PLAN

The experiment was conducted in two phases; first in external winter conditions and secondly under controlled conditions of −5°C. The external temperatures during the casting and curing period are listed in Table 3. The casting of specimen was done during the latter part of the day or preferably during the night time, to fully utilize the sub-zero temperature conditions. Samples cast under controlled conditions were stored in a freezer at −5°C, within 30 minutes of mixing. The samples were stored at −5C° for the first 2 days and after that were exposed to normal weather conditions.

Date	Max. daily temp.(C)	Min. daily temp.(C)	Date	Max. daily temp.(C)	Min. daily temp.(C)	Date	Max. daily temp.(C)	Min. daily temp.(C)
15-01-2016 (Casting 1)	6	−4	**29-01-2016 (Casting 2)**	7	−2	12-02-2016	9	−1
16-01-2016	8	−3	30-01-2016	10	2	13-02-2016	8	−2
17-01-2016	10	−4	31-01-2016	10	2	14-02-2016	8	0
18-01-2016	11	−3	01-02-2016	7	−2	**15-02-2016 (Casting 5)**	6	−3
19-01-2016	10	−3	02-02-2016	8	−2	16-02-2016	11	−1
20-01-2016	8	−2	03-02-2016	8	−3	17-02-2016	11	−1
21-01-2016	5	−4	04-02-2016	9	−2	18-02-2016	12	0
22-01-2016	6	−5	05-02-2016	5	−3	19-02-2016	10	−2
23-01-2016	6	−4	**06-02-2016 (Casting 3)**	5	−4	20-02-2016	13	−2
24-01-2016	8	−2	07-02-2016	7	−3	21-02-2016	12	−1
25-01-2016	5	−4	08-02-2016	10	−1	22-02-2016	14	2
26-01-2016	5	−5	09-02-2016	11	−2	23-02-2016	13	1
27-01-2016	7	−2	**10-02-2016 (Casting 4)**	5	−2	24-02-2016	11	2
28-01-2016	9	−1	11-02-2016	7	−3	25-02-2016	15	4

Table 3.1 Exterior winter conditions

CHAPTER 4

MATERIALS AND METHODS

4.1 MATERIALS

The various materials required for this project work include Cement (Khyber®) of grade 53 conforming to IS 12269 - 1987, fine aggregate conforming to IS 383 1987. The fine aggregate has been collected from Ganderbal (Jammu & Kashmir, India) and coarse aggregate conforming to IS 383 1987 also collected from Ganderbal. The origin of both the fine and coarse aggregates are being river Sand and gravel. Concrete is prepared by mixing various constituents like cement, aggregates, water etc. which are economically available. Ordinary Portland cement of grade 53 will be used throughout the work. The fine aggregate to be used in this investigation is clean river sand, whose maximum size is 4.75 mm, conforming to grading zone II. Machine crushed stone angular in shape will be used as coarse aggregate. Two sizes of coarse aggregate are used; one 10 mm and other 20mm. Sodium nitrite and Potassium carbonate are to be used as antifreeze admixtures to study the various properties of concrete produced in external winter conditions of Kashmir.

S.No	Material Property	Value
1.	Initial setting time of cement	3 hr 40 min
2.	Final setting time of cement	5 hr 20 min
3.	Specific gravity of cement	3.12
4.	Specific gravity of sand	2.63
5.	Specific gravity of coarse aggregate	2.67
6.	Standard consistency of cement	30%
7.	Fineness modulus of sand	2.58

8.	Design Mix	1:1.35:2.87

Table 4.1 Various material properties

4.2 CASTING OF TEST SPECIMEN (AS PER IS: 516-1959)

4.2.1 PREPARATION OF MATERIALS

All materials shall be brought to room temperature, preferably 27 ± 3 °C before commencing the results. The cement samples, on arrival at the laboratory, shall be thoroughly mixed dry either by hand or in a suitable mixer in such a manner as to ensure the greatest possible blending and uniformity in the material, care is being taken to avoid the intrusion of foreign matter. The cement shall then be stored in a dry place, preferably in air-tight metal containers. Samples of aggregates for each batch of concrete shall be of the desired grading and shall be in an air-dried condition. In general, the aggregate shall be separated into fine and coarse fraction and recombined for each concrete batch in such a manner as to produce the desired grading. IS sieve 480 shall be normally used for separating the fine and coarse fractions, but where special grading's are being investigated, both fine and coarse fractions shall be further separated into different sizes.

4.2.2 PROPORTIONING

The proportions of the materials, including water, in concrete mixes used for determining the suitability of the materials available, shall be similar in all respects to those to be employed in the work. Where the proportions of the ingredients of the concrete as used on the site are to be specified by volume, they shall be calculated from the proportions by weight used in the test cubes and the unit weights of the materials.

4.2.3 MIXING CONCRETE (HAND MIXING)

The concrete shall be mixed by hand or preferably in a laboratory batch mixer, in such a manner as to avoid loss of water or other materials. Each batch of concrete shall be of such a size as to leave about 10 percent excess after casting the desired number of test specimens.

The concrete batch shall be mixed on a water-tight, non-absorbent platform with a shovel, trowel or similar suitable implement, using the following procedure:

- The cement and fine aggregate shall be mixed dry until the mixture is thoroughly blended and is uniform in color.

- The coarse aggregate shall then be added and mixed with the cement and fine aggregate until the coarse aggregate is uniformly distributed throughout the batch, and

- The water shall then be added and the entire batch mixed until the concrete appears to be homogenous and has the desired consistency. If repeated mixing is necessary, because of the addition of water in increments while adjusting the consistency, the batch shall be discarded and a fresh batch made without interrupting the mixing to make trial consistency tests.

4.2.4 COMPACTION OF TEST SPECIMENS: (AS PER IS: 516-1959)

The test specimens shall be made as soon as practicable after mixing, and in such a way as to produce full compaction of the concrete with neither segregation nor excessive laitance. The concrete shall be filled into the mould in layers approximately 5 cm deep.

In placing each scoopful of concrete, the scoop shall be moved around the top edge of the mould as the concrete slides from it, in order to ensure a symmetrical distribution of the concrete within the mould. Each layer shall be compacted either by hand or by vibration as described below. After the top layer has been compacted, the surface of the concrete shall be finished level with the top of the mould, using a trowel, and covered with a glass or metal plate to prevent evaporation.

When compacting by vibration, each layer shall be vibrated by means of an electric or pneumatic hammer or vibrator or by means of a suitable vibrating table until the specified condition is attained.

4.2.5 PLACING MOULDS ON THE VIBRATING TABLE

The moulds may be rigidly clamped to the vibrating table in such a manner that they have contact with the support in as many and in the most suitable places possible, so that the vibration amplitude is fairly uniform over the whole range of the support and moulds. With the rigid and uniform clamping of the moulds the frequency and amplitude of vibration of the table are uniformly transmitted to the mould as well as the fresh concrete. For smaller units when the moulds are not rigidly clamped on the table, they are repeatedly thrown into the air in a haphazard manner owing to the vibration acceleration of the tables, which is generally considerably greater than the acceleration due to gravity. During this process the concrete may be subjected to impacts with quite high acceleration but there may be considerable loss of energy transmitted to the concrete and there may be damage to the concrete, moulds and table.

4.2.6 CURING OF TEST SPECIMENS: (AS PER IS: 516-1959)

The test specimens shall be stored on the site at a place free from vibration, under damp matting, sacks or other similar material for 24 hours + ½ hour from the time of adding the water to the other ingredients as shown in Figure . After the period of 24 hours, they shall be marked for later identification, removed from the moulds and, unless required for testing within 24 hours, stored in clean water until they are transported to the testing laboratory. They shall be sent to the testing laboratory well packed in damp sand, damp sacks, or other suitable material so as to arrive there in a damp condition not less than 24 hours before the time of test. On arrival at the testing laboratory, the specimens shall be stored in water until the time of test.

Fig 4.1 Curing of specimens

4.3 METHODS

4.3.1 COMPRESSIVE STRENGTH

For cube compression testing of concrete, 150mm cubes were used. All the cubes were tested in saturated condition, after wiping out the surface moisture. The tests were carried out after the specimen has been centered in the testing machine. Loading was continued till the specimen fails and reading note down from the automatic universal testing machine. The ultimate load divided by the cross sectional area of the specimen is equal to the ultimate cube compressive strength.

$$f_c = P/A$$

Where,

f_c = compressive strength in MPa

P= load in Newton

A= area of the specimen in mm

Fig 4.2 Compressive strength test

4.3.2 SPLIT TENSILE STRENGTH

This is an indirect test to determine the tensile strength of cylindrical specimens. Splitting tensile strength tests were carried out on cylinder specimens of size 150 mm diameter and 300 mm length. To avoid the direct load on the specimen the cylindrical specimens were kept below the iron plates .The load was applied gradually till the specimens split and readings were noted. The test set up for the splitting tensile strength on the cylinder specimen, with the iron plates to avoid the direct load on the specimen is shown in Figure. Patterns of typical splitting tensile failure mode shapes of HPC cylinder specimens are shown in Figure. The splitting tensile strength has been calculated using the following equation.

$$\mathbf{f_t = 2P/\pi\ DL}$$

Where

f_t = splitting tensile strength of the specimen in MPa

P = maximum load in N applied to the specimen

D = measured diameter of the specimen in mm, and

L = measured length of the specimen in mm.

Fig 4.3 Split Tensile Strength Test

4.3.3 FLEXURAL STRENGTH

Flexural strength tests were carried out on 100mm x 100 mm x 500 mm HPC prisms by subjecting the specimen to two-point loading to determine the flexural strength.

The test set up for the flexural strength on the prism specimen with the necessary settings is shown in Figure. The flexural strength or modulus of rupture has been calculated using the following formula:

$$\mathbf{f_r = PL/BD^2}$$

Where

fr = flexural tensile strength of the specimen in MPa

P = maximum load in N applied to the specimen

L = length in mm of the span on which the specimen was supported

B = measured width of the specimen in mm, and

D = measured depth of the specimen in mm at the point of failure

Fig 4.4 Flexural Strength Test

4.3.4 STATIC MODULUS OF ELASTICITY AND POISSON'S RATIO

This test method covers determination of (1) chord modulus of elasticity (Young"s) and (2) Poisson"s ratio of molded concrete cylinders and diamond-drilled concrete cores when under longitudinal compressive stress. The modulus of elasticity and Poisson"s ratio values, applicable within the customary working stress range (0 to 40 % of ultimate concrete strength), are used in sizing of reinforced and non-reinforced structural members, establishing the quantity of reinforcement, and computing stress for observed strains. The modulus of elasticity values obtained will usually be less than moduli derived under rapid load application (dynamic or seismic rates, for example), and will usually be greater than values under slow load application or extended load duration, given other test conditions being the same.
Procedure: Maintain the ambient temperature and humidity as constant as possible throughout the test. Record any unusual fluctuation in temperature or humidity in the report. Use companion specimens to determine the compressive strength in accordance with Test Method C 39/C 39M prior to the test for modulus of elasticity. Place the specimen, with the strain-measuring equipment attached, on the lower platen or bearing block of the testing machine. Carefully align the axis of the specimen with the center of thrust of the spherically-seated upper bearing block. Note the reading on the strain indicators. As the spherically-seated block is brought slowly to bear upon the specimen, rotate the movable portion of the block gently by hand so that uniform seating is obtained. Load the specimen at least twice. Do not record any data during the first loading. Base calculations are on the average of the results of the subsequent loadings.

During the first loading, which is primarily for the seating of the gages, observe the performance of the gages and correct any unusual behavior prior to the second loading. Obtain each set of readings as follows: Apply the load continuously and without shock. Set testing machines of the screw type so that the moving head travels at a rate of about 0.05 in. (1.25 mm)/min when the machine is running idle. Record, without interruption of loading, the applied load and longitudinal strain at the point (1) when the longitudinal strain is 50 millionths and (2) when the applied load is equal to 40 % of the ultimate load. Longitudinal strain is defined as the total longitudinal deformation divided by the effective gage length. If Poisson"s ratio is to be determined, record the transverse strain at the same points. If a stress-strain curve is to be determined, take readings at two or more intermediate points without interruption of loading; or use an instrument that makes a continuous record. Immediately upon reaching the maximum load, except on the final loading, reduce the load to zero at the same rate at which it was applied. If the observer fails to obtain a reading, complete the loading cycle and then repeat it.

Fig 4.5 Cylindrical specimen fitted with a compressometer extensometer arrangement

Fig 4.6 Arrangement for measuring elastic modulus and Poisson's ratio

Calculate the modulus of elasticity, to the nearest 50 000 psi (344.74 MPa) as fllows

$$E = \frac{S_2 - S_1}{\varepsilon_2 - 0.000050}$$

Calculate Poisson"s ratio, to the nearest 0.01, as follows:

$$\mu = \frac{\varepsilon_{t2} - \varepsilon_{t1}}{\varepsilon_2 - 0.000050}$$

Where:

μ = Poisson"s ratio,

ε_{t2} = transverse strain at midheight of the specimen produced by stress S2, and

ε_{t1} = transverse strain at midheight of the specimen produced by stress S1.

CHAPTER 5

TEST RESULTS

5.1 COMPRESSIVE STRENGTH

The cube compressive strength results at the various ages 7 and 28 days and at the different admixtures percentages as shown in Fig 7, 8, 9 and 10. The compressive strength values decreases with decrease in temperature which the concrete is exposed. The maximum decrease of the compressive strength observed under controlled concrete (-5 °C).

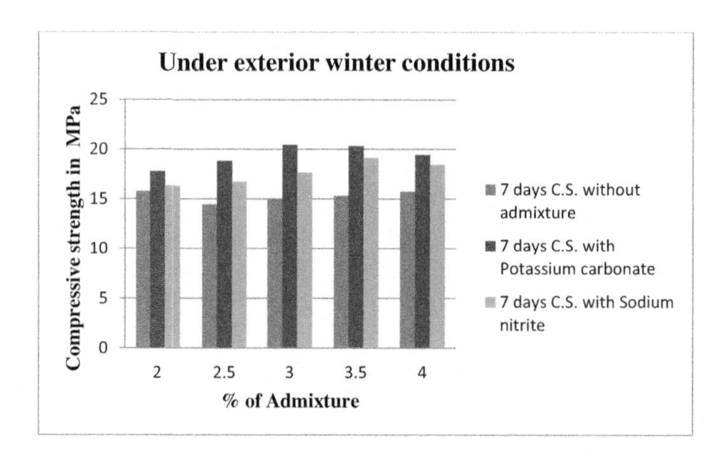

Fig 5.1 Compressive strength at 7 days under exterior winter condition

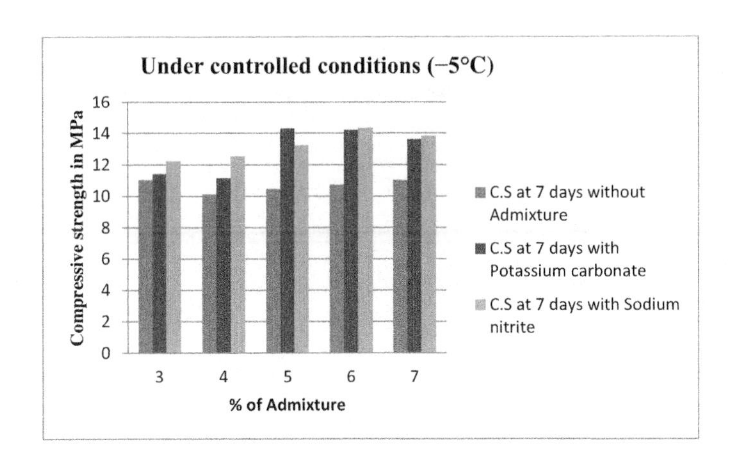

Fig 5.2 Compressive strength at 7 days under controlled condition

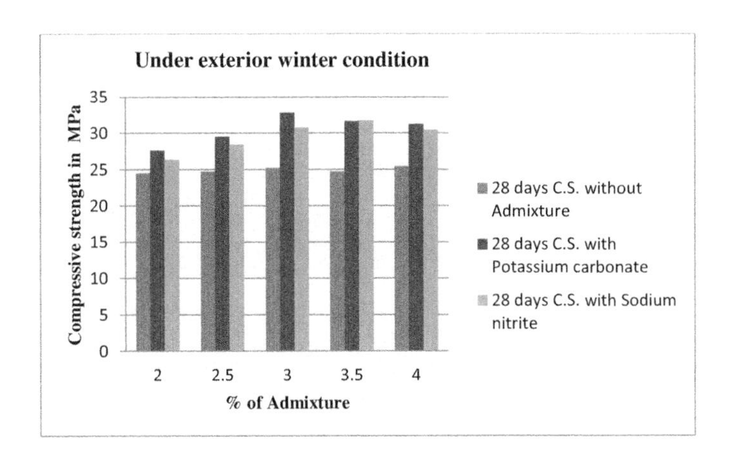

Fig 5.3 Compressive strength at 28 days under exterior winter condition

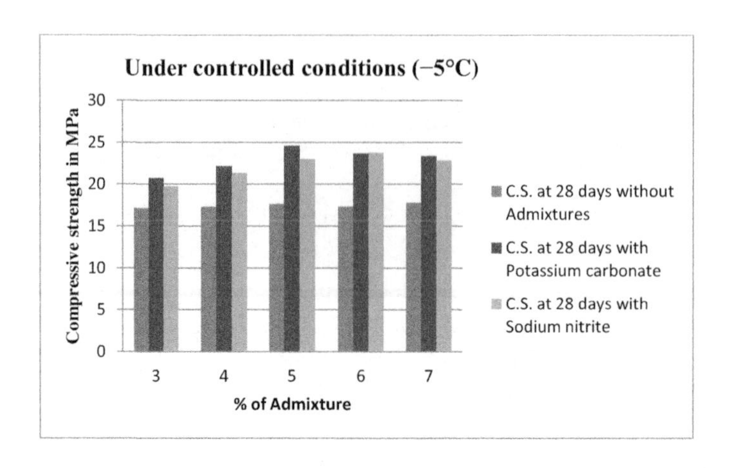

Fig 5.4 Compressive strength at 28 days under controlled condition

5.2 SPLIT TENSILE STRENGTH

The cylinder splitting tensile strength of results at the various ages 7 and 28 day at the different admixtures percentages as shown in Fig 11, 12, 13 and 14. The splitting tensile strength values decreases with decrease in temperature which the exposed concrete. The maximum decrease of the split tensile strength observed at decrease temperature of -5 °C.

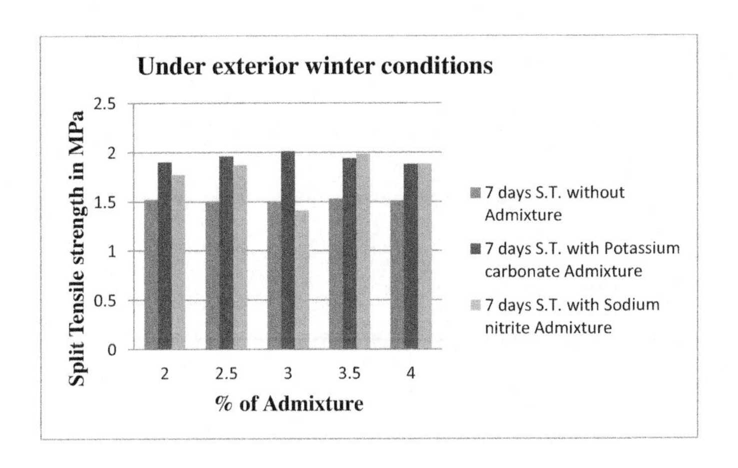

Fig 5.5 Splitting tensile strength at 7 days under exterior winter condition

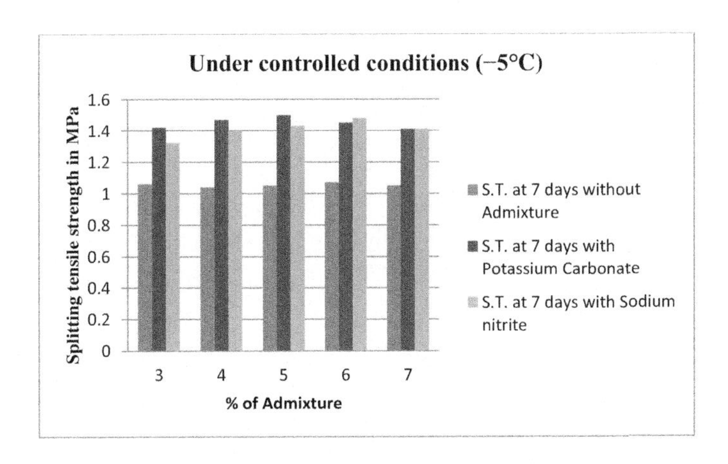

Fig 5.6 Splitting tensile strength at 7 days under controlled condition

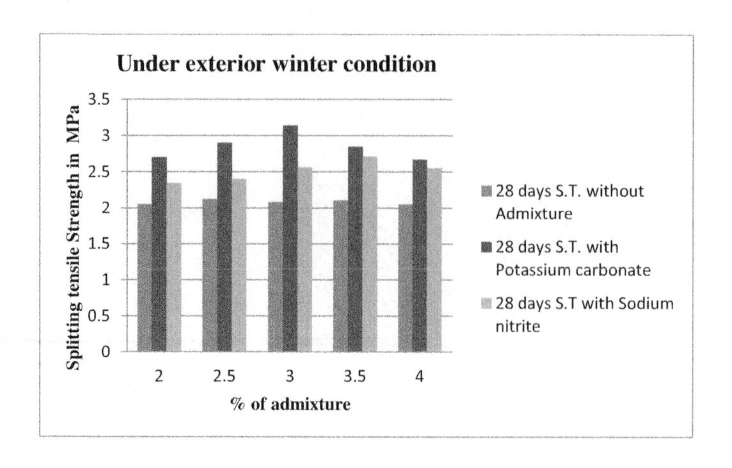

Fig 5.7 Splitting tensile strength at 28 days under exterior winter condition

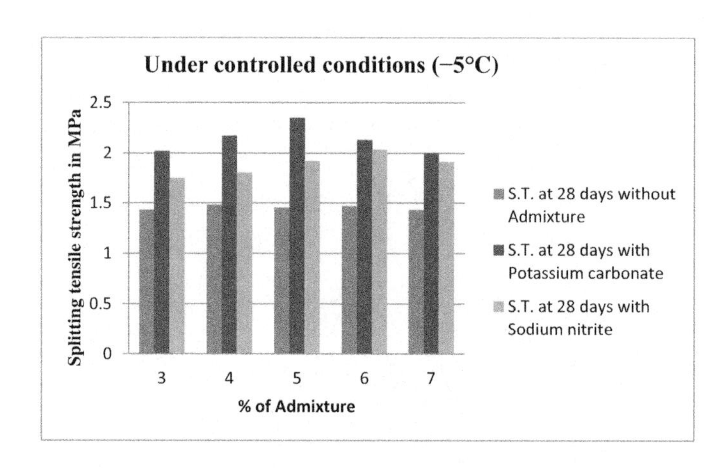

Fig 5.8 Splitting tensile strength at 28 days under controlled condition

5.3 FLEXURAL STRENGTH

The flexural strength of concrete results at the various ages 7 and 28 day at the different admixtures percentages as shown in Fig 15, 16, 17 and 18. The flexural strength values decreases with decrease in temperature which the concrete is exposed. The maximum decrease of the flexural strength observed at decrease temperature of -5 °C.

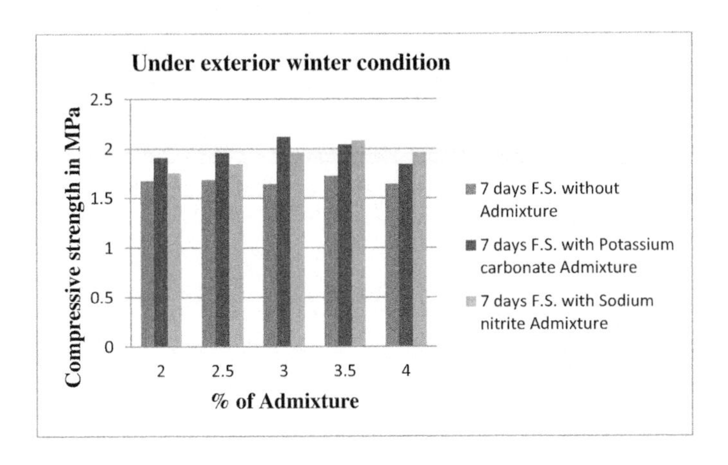

Fig 5.9 Flexural strength at 7 days under exterior winter condition

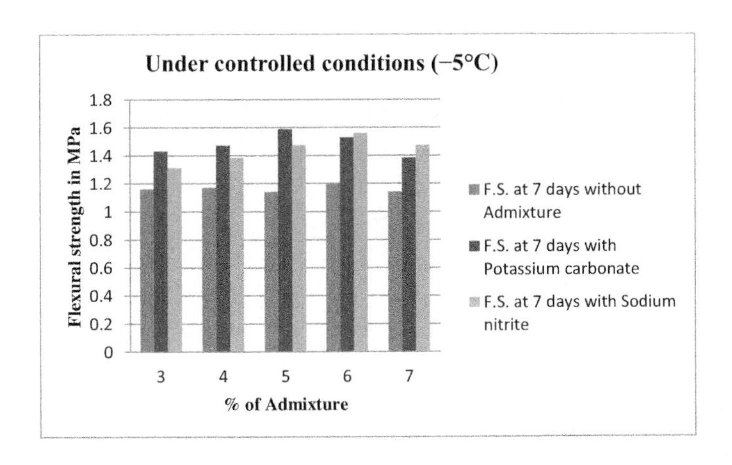

Fig 5.10 Flexural strength at 7 days under controlled condition

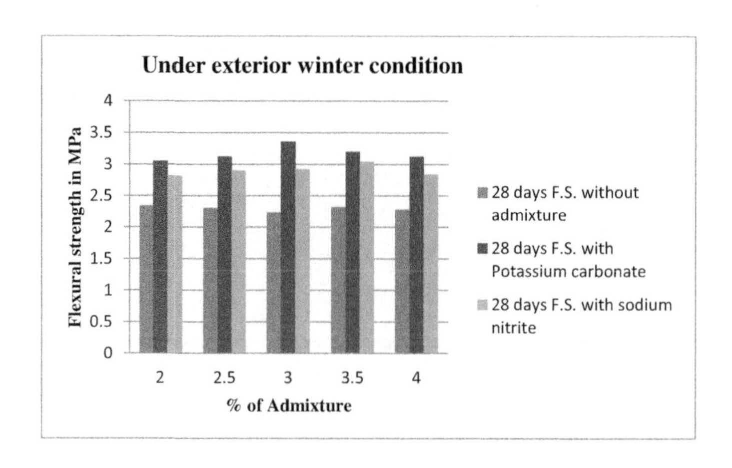

Fig 5.11 Flexural strength at 28 days under exterior winter condition

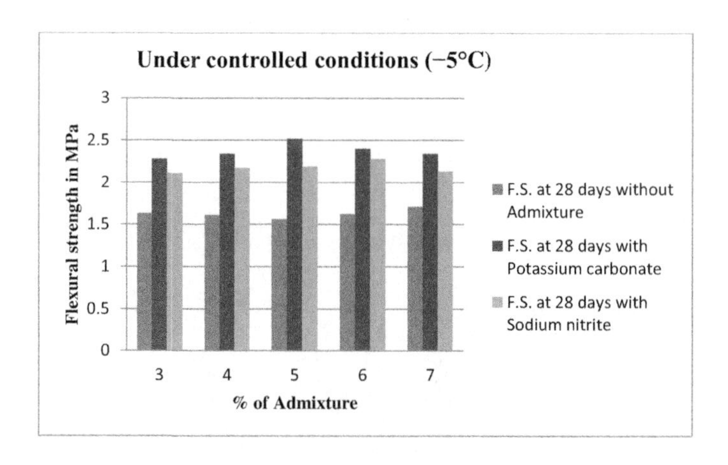

Fig 5.12 Flexural strength at 28 days under controlled condition

5.4 STATIC MODULUS OF ELASTICITY

The static modulus of elasticity results at 28 day at the different admixtures percentages as shown in Fig 19 & 20. The static modulus of elasticity values decreases with decrease in temperature which the exposed concrete. The maximum decrease of the modulus of elasticity observed at decrease temperature of -5 °C.

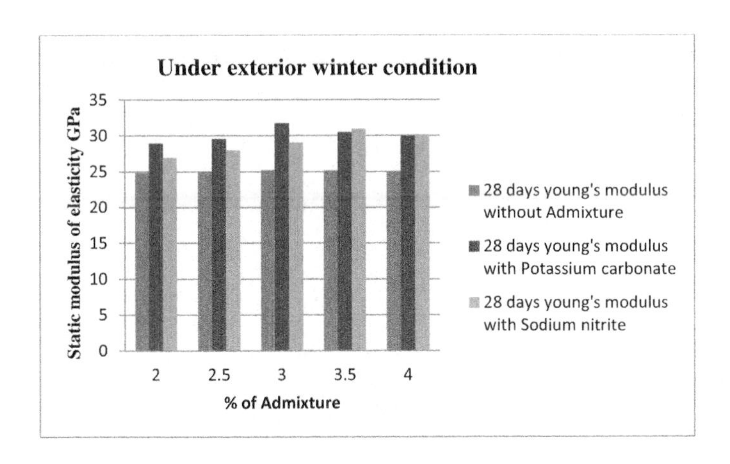

Fig 5.13 Static modulus of elasticity at 28 days under exterior winter condition

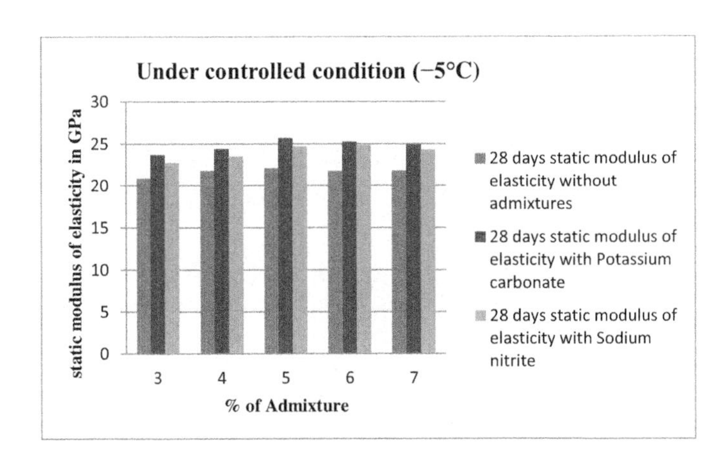

Fig 5.14 Static modulus of elasticity at 28 days under controlled condition

5.5 POISSON'S RATIO

The Poisson's ratio of concrete results at 28 day at the different admixtures percentages as shown in Fig 21 & 22. The Poisson's ratio values decreases with decrease in temperature which the concrete is exposed. The maximum decrease of the Poisson's ratio observed at decrease temperature of -5 °C.

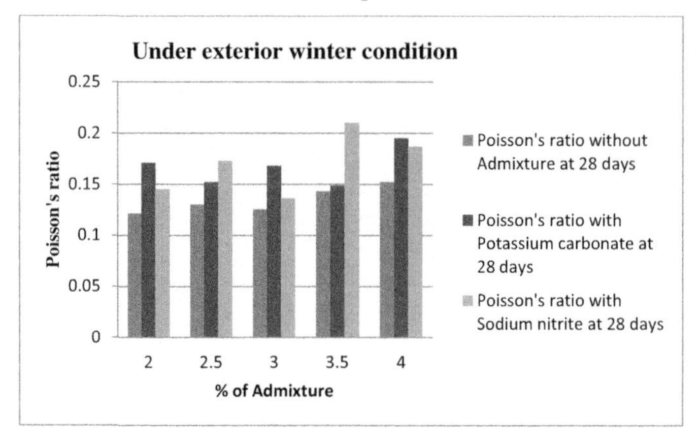

Fig 5.15 Poisson's ratio at 28 days under exterior winter condition

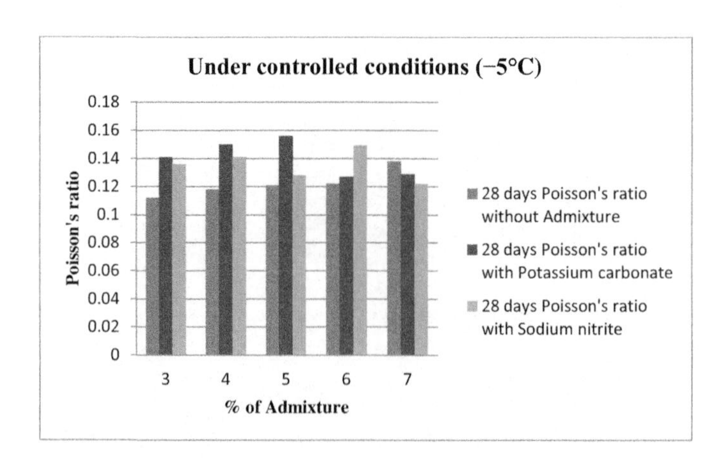

Fig 5.16 Poisson's ratio at 28 days under controlled condition

CHAPTER 6

CONCLUSION

- For exterior weather conditions (as stated in Table 3) maximum values of concrete strength viz. Compressive strength, Flexural strength and Split tensile strength were observed for 3% K_2CO_3 and 3.5% $NaNO_2$. For 3% K_2CO_3 the 28-day compressive strength showed a 30% increase while the Split tensile strength and Flexural strength showed an increase of 51% and 50% respectively. For 3.5% $NaNO_2$ the 28-day compressive strength showed a 28% increase while the Split tensile strength and Flexural strength showed an increase of 29% and 31% respectively.
- The increase in 3-day and 7-day compressive strength is higher than the increase in 28- day compressive strength. Thus increase in early strength is more significant than increase in strength later on. The increase in early strength shows that the freezing of fresh concrete did not take place.
- For controlled conditions (wherein the concrete samples were exposed to a temperature of −5°C for first two days) maximum values of concrete strength viz. Compressive strength, Flexural strength and Split tensile strength were observed for 5% K_2CO_3 and 6% $NaNO_2$. For 5% K_2CO_3 the 28-day compressive strength showed a 40% increase while the Split tensile strength and Flexural strength showed an increase of 62% and 62% respectively. For 6% $NaNO_2$ the 28-day compressive strength showed a 37% increase while the Split tensile strength and Flexural strength showed an increase of 38% and 41% respectively.
- The moduli of elasticity (E) values are in good agreement with the compressive strength values, with the maximum value of E corresponding to the maximum compressive strength. For K_2CO_3 the value of E showed an increase of 25.86%, while for $NaNO_2$ the value of e showed an increase of

22.14% at 3% and 3.5% respectively.

- The poisson's ratio however did not seem to follow any specific trends. This may be attributed to the fact that the determination of poisson's ratio may be more sensitive to the loading rate of the arrangement than modulus of elasticity (the provisions of the code specify a certain rate of loading which couldn't be provided due to certain instrumental restrictions). This however is only a speculation and needs further study.
- One of the key observations is that in exterior weather conditions the design strength at optimum dosages was achieved which is not the case with controlled conditions. Thus it can be said that simple compounds or their combinations can be used for the desired effects in the given conditions. Complex formulations of admixtures, which are quite expensive, may not be needed.

REFERENCES

- Fatma Karagol Ramazan Demirboga Waleed H. Khushefati " Behavior of fresh and hardened concretes with antifreeze admixtures in deep-freeze low temperatures and exterior winter conditions". Construction and Building Materials 76 (2015) 388–395.
- Mustafa Cullu, Metin Arslan "The effects of antifreeze use on physical and mechanical properties of concrete produced in cold weather" Composites: Part B 50 (2013) 202–209.
- Shetty, M.S. Concrete Technology Theory and Practice (2011). S.Chand and Company LTD.
- Joseph F. Lamond and James H. Pielert, Editors Significance of Tests and Properties of Concrete and Concrete-Making Materials (2006) STP 169D.
- Tikalsky (November 2005) "Concrete Admixtures that Defend Against Salt Scaling and Freeze Thaw" Cold Regions Research and Engineering Laboratory.
- Korhonen C.J. and Semen P.M. (April 2005) "Placing Antifreeze Concrete at Grand Forks Air Force Base" Cold Regions Research and Engineering Laboratory, Technical Report 05-9
- Korhonen C.J., Semen P.M., Barna L.A. (February 2004) "Extending the Season for Concrete Construction and Repair Phase 1 – Establishing the Technology" Cold Regions Research and Engineering Laboratory, Techinical Report 04-02
- ACI 306 (2002) Cold Weather Concreting ACI 306R – 02, ACI Committee306 Detroit, Michigan: American Concrete Institute.
- Roger Rixom and Noel Mailvaganam, Chemical Admixtures for Concrete third edition (1999).
- Korhonen C.J., Cortez R.E., Duning T.A. and Jeknavorian A.A. (October

1997) "Antifreeze Admixtures for Concrete" Cold Regions Research and Engineering Laboratory, Special Report 97-26.

- Charles J. Korhonen Antifreeze Admixtures for Cold Regions Concreting A Literature Review September 1990.
- Kukko, H. and I. Koskinen (1988) RILEM recommendations for concreting on cold weather. Technical Research Centre of Finland (VTT), Concrete and Silicate Laboratory, Research Note 827.
- ACI (1988) Cold weather concreting. ACI 306R-88. Detroit, Michigan: American Concrete Institute.
- Kivekaes L., Huovinen S. Leivo M. (1985) "Concrete under Arctic Conditions", Technical Research Center, Finland.
- Kuzmin, Y. D. (1976) Concretes with Antifreeze Admixtures. Kiev: Budivelnik Publishing House.

Authors

Panga Narasimha Reddy

Mr. Panga Narasimha Reddy is currently Pursuing PhD at National Institute of Technology, Srinagar (Jammu & Kashmir). He received his B.Tech (Civil Engineering) and M.Tech (Structural Engineering) from Jawaharlal Nehru Technological University, Anantapur.

Javed Ahmed Naqash

Dr. Javed Ahmed Naqash is working as Associate Professor at Department of Civil Engineering, National Institute of Technology, Srinagar. He received his B.Tech (Civil Engineering) from REC Srinagar and M.Tech (Structural Engineering) from IIT Delhi. He received a doctorate in the field of Seismic microzonation and Concrete structures from IIT Roorkee.

Avuthu Narender Reddy

Mr. Avuthu Narender Reddy is currently Pursuing PhD from Vellore Institute of Technology, Vellore. He received his M.Tech (Structural Engineering) from

Jawaharlal Nehru Technological University, Kakinada and B.Tech (Civil Engineering) from Acharya Nagarjuna University, Guntur. He has published number of research articles in reputed journals and he is contributing as Editorial Board Member for 3 International journals and reviewer for more than 7 International journals

Bode Venkata KavyaTeja

Mrs. Bode venkata kavyateja is currently Pursuing PhD from Jawaharlal Nehru Technological University, Anantapur. She received her B.Tech (Civil Engineering) and M.Tech (Structural Engineering) from Jawaharlal Nehru Technological University, Anantapur. She has published number of research articles in reputed journals.

Printed in Great Britain
by Amazon

87127709R00037